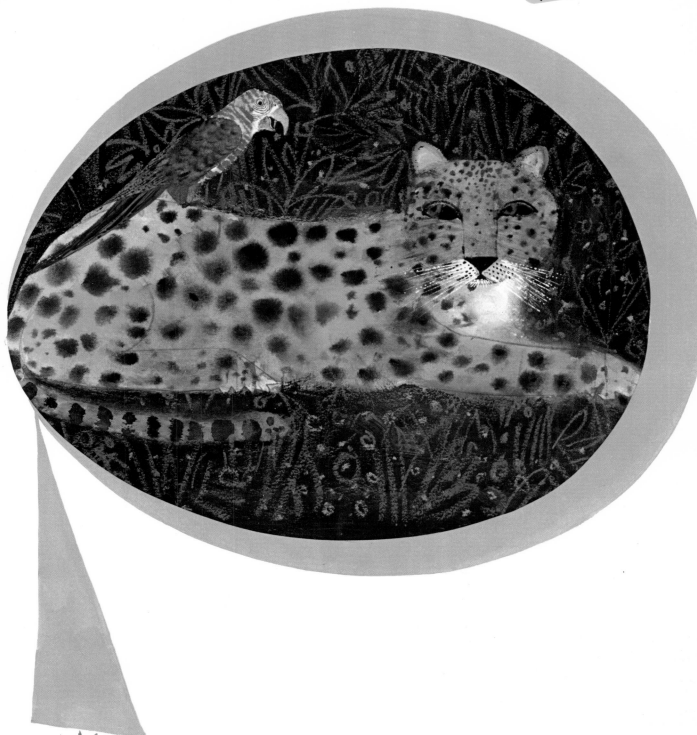

Oxford University Press

OXFORD TORONTO MELBOURNE

Brian Wildsmith

What the Moon saw

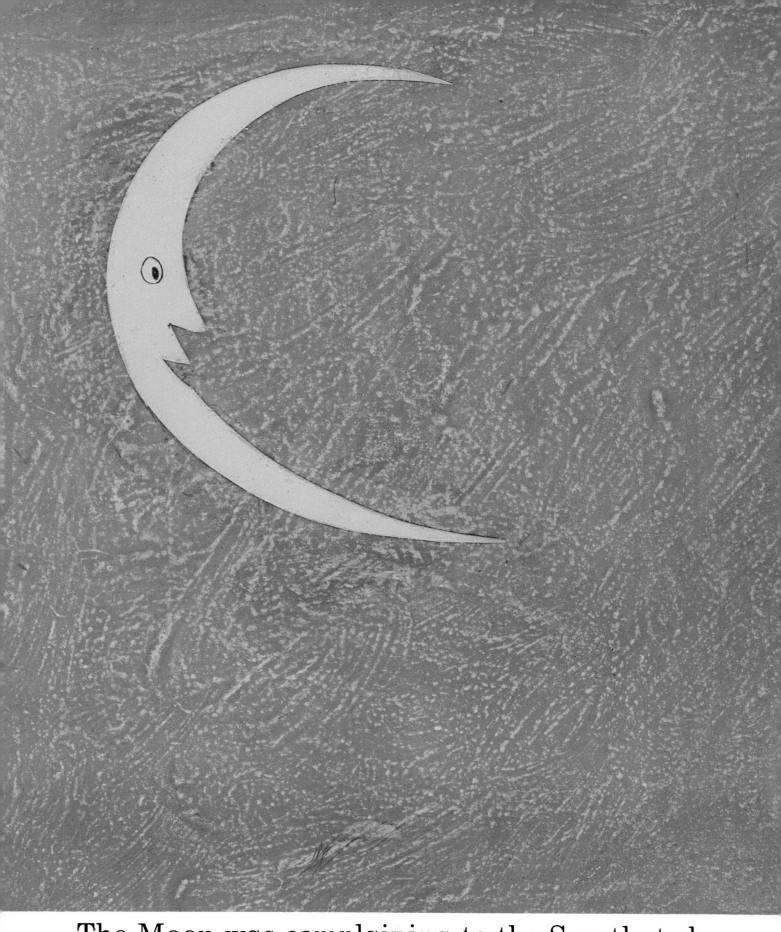

The Moon was complaining to the Sun that she had never really seen the world that lay below.

'I,' said the Sun, 'have seen everything, and I will show you some of the many things to be found there.'

'That is a city. It has **many** houses and buildings.'

'There is a village. It has **few** houses.'

'This is the **outside** of a house,

and if you peep through the windows
you can see the **inside**.'

'Look further and you will see a **big** forest,

and in it is a **little** flower.'

'Here we look at a dog from the **front**,

and here from the **back**.'

'This elephant is a **heavy** beast,

but this bird is very **light**.'

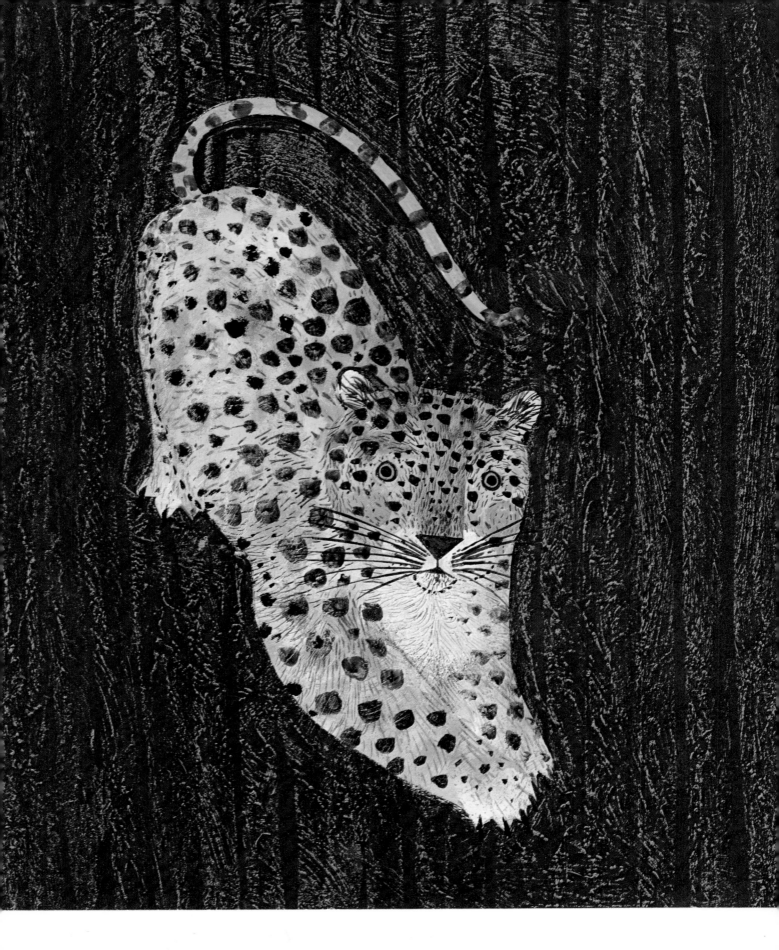

'The leopard's coat is **patterned** with spots,

the mane of the lion is **plain**.'

'Here is a **fat** hippopotamus,

and there is a **thin** lizard.'

'A **feathered** pelican,

and a **furry** llama.'

'The giraffe's neck is **long**,

but the racoon has a **short** neck.'

'Look at the **fierce** tiger,

and the **timid** rabbit.'

'A **weak** kitten,

a **strong** bear.'

'When running the cheetah is **fast**,

but the tortoise is always **slow**.'

'Isn't that all wonderful to see,' said the Sun,

'how lucky I am!
I believe there is nothing I haven't seen.'

'Yes, there is,' answered the Moon. 'Every night
I see something which you have never seen,
nor ever will – the **dark**.'